Late Fall Bucolics

Anne Coray

for Anne
Also with an "e" –
Wishing you continued
success with your
fine writing.
Anne Coray

A Publication of The Poetry Box®

Editing & Book Design by Shawn Aveningo Sanders
Cover Design by Shawn Aveningo Sanders
Cover Painting by Alan Scales via Unsplash

ISBN: 978-1-956285-11-6
Printed in the United States of America.
Wholesale distribution via Ingram.

Published by The Poetry Box®, July 2022
Portland, Oregon
https://ThePoetryBox.com

A cow gave birth to a fire; she wanted to lick it, but it burned; she wanted to leave it, but she could not because it was her own child.

—Ethiopian Proverb

Fire is never a gentle master.

—Traditional Proverb

Contents

Brazen Heat

Five thousand feet against cobalt sky
The consummate white is maiden snow
Shouldering mountaintops' dip and rise
When the day is clear and the lake below

Reflects the waning sun's full glint.
We are finished with fire. The brazen heat
Of a cracked July has fled, spent
Except in memory, along with smoke from forest

Charred to north and west, that blocked our flight.
But what, in weeks to come, shall we recall?
Our trust's in balance, a means to set things right.
Thus: we swam. We dove. We tanned. O memory! Poor tool

To hold against the sum of the planet's warming.
Like stubborn leaf to barren branch, to this winter scene, we cling.

Ruse

Berries of the mountain ash still cling
To the leafless tree: crimson beads in clusters
Plump against thin stalks of tawny grasses. Nothing
Strange in this, a canvas known to mid-November,

But the ground, unfrozen, weeps at every print,
The boot heels' rubber stuccoed with sodden dirt.
Gone our five-day stretch of lucid weather; hints
Of normalcy have turned to ruse. A covert

Painter laid the mountaintops with snow then loaded
His brush for us with rain. Storm after storm
Has rent our beach, while northern coasts erode.
But, as luck or whim will have it, no harm

Comes to us—yet. And if the mad artist should efface
Our portraits, we hope something remains of place.

Apprenticed

Hard to know what it means anymore, a "sense of place."
Yet that was the poet's instruction, to write
That which we knew. Now this season bears little trace
To the autumn of my youth; only the light's

Diminishment is familiar. Ice pockets on creeks,
Hoar frost stippling deck and willow, the air
Newly honed; ghosted breath, reddened cheek:
These common descriptives no longer

Greet my pen. Do I miss them? Yes—though my hand's
Been hired out: I am apprenticed
To the painter, crazed with work. The land
He claims must be transformed. I am complicit.

So the red line climbs behind the glass:
Forty-nine, and glaciers dwindle in our mountain pass.

Veteran Creator

He offers me a palette; I do not pass
It back. Begin, he says, with madder brown
For alder. Brush in hand, I bridge a deep crevasse,
Then move upslope. He nods, and leaves me on my own.

I am no painter. The bristles of my tool are stiff
As I slash out trunk and branch, but soon
My skill is sharpened, and I become my myth.
Master, maker, veteran creator… the loon

Laughs crazily, they say, but what of us? Black
Mars will be our final color, if we cannot halt
This growth … a tangled patch consuming track
Of caribou; Brush, brush, crosshatch, quicken, the fault's

Not mine, I did not sign. Surely I was coerced.
Ho ho. Every stroke's in oil, and I'm immersed.

First Infraction

Saturated air, the pass immersed
In fog. We still can't fly, although the rain
This day has ceased its drumming. A burst
Of sun emblazons one wing of our plane

That sits on rocky beach, tied down, fueled
Up. What a lovely color: its body a warm
And deep maroon, with accents black. The cool
Steel frame in my palm makes me forget the harm

The engine causes, though it's slim compared
To the silver raptors mining the skies
Above, en route to Russia. Who dared
This first infraction? The brothers, yes, but the painter's eyes

Are frightening. He spots a jet, and deems the contrail
Beautiful. One lung of oxygen for sale.

Cast in Steel

Bristol Bay fishing boats no longer manned by sail.
The year is 1951; soon, in every village,
Airplanes displace dog teams to deliver mail.
Now four-wheelers furrow tundra and ridge,

Jet boats and hovercraft scour rivers.
We study the cost, but access is our idol,
Cast in steel. The craftsman knows his art. Sculptor,
Half-brother, reaper of ore, your tools

Are drill and welder. You've fastened us together.
I shall paint you into the frame, an outdoor portrait:
White catkins for a backdrop, November
Willows posing for spring. Already the date

Means nothing. Thirty or one. December will merge with March
To—what might we skip, from Easter to Halloween? The scorch.

Her Loud Red Coat

The threat of fire. The smoke. The sear and scorch.
And that is why Jane, in her loud red coat
Marches on Washington: a resolute torch
Ignited by duty, her best leading part

Since '78. A kind of homecoming for her,
Protest of a different war: this one global,
Decades later. She's 82, not as sure
In her step, but prudent and ever beautiful,

Willing to face arrest. Handcuffed, jailed,
Imperturbable, quick with her smile—
She's the perfect model, but my mentor has prevailed:
He's unhappy with my portraits. My style's

Best suited for landscape; I cannot honor this woman
In paint. Commemoration is not the plan.

Orange and Vermilion

Incensed, debased, I venture: What is the plan?
The painter laughs insanely. *I think you know it.*
I move to my studio, draw the curtain.
Whose ailment, now? Cowed by the Master's fixed

Ambition, not daring to defy his path.
His haughty stare, his tireless hand, have sent me
To my easel, forced me to arrest my wrath.
I am incapable of refusal. I'm at his mercy.

But I'm no servant!—Except to movement,
Indentured to wheel and cog. Thus with turpentine
And rag I scrub my latest canvas, remove all hint
Of evergreen. It's on to orange and vermilion.

So long! All will burn, but how magnificent the color.
The contract is invisible. A system of unfree labor.

Spell of Pigment

There's no compensation for this labor,
Not a single dollar, no plot of land. Instead,
Destruction. I could nick my fingers or sever
My wrist, like the slave who refused

To cut cane. What good? My mechanical
Hand would still grip the palette knife, my favorite
New applicator. Layer on layer, the spell
Of pigment upon me: rose and cobalt violet,

Undertones to scrape down and recoat
With opaque shades of white and pewter.
It adds complexity. And here, a wash, transparent
Blue for a bit of lake below the fog, pure

Water in the zone of a world that's no longer stable.
Only the fog today is motionless, still as stones or wool.

Yarn

Mittens, sweaters, insoles for boots, old blankets: wool
Was our protector, but sometimes it made us itch.
A fabric lost to my closet, as the simple
"Knit one, purl two" I learned as a child is a niche

For artisans only. From drop spindle to ring
And rotor, the yarn is spun—faster is better—
A burst of textiles, ushered in with the burning
Of coal; the blackened country, child labor,

And chimney sweeps—young boys, five or eight
Years of age—heads shaved so they wouldn't catch
Fire, the soot that infused their lungs and plagued
Their skin with cancer. Low cost. Dispatch

To the 21st century, our firmament trapped
With emissions, and we, in our modern blanket, wrapped.

Rapt

Evening, the welcome dark. We lie on the bed, rapt
By fire, visible through the glass door of our stove.
Flame, a serpent's tongue, coals hot coral, capped
With beetle-killed rounds of spruce. The insects love

The changing climate, as Steve and I
Love wood: the texture, the heft, the harvest,
The sawing and hauling, until the split dry
Log, in its smelted coffin, is laid to rest.

And mostly the warmth, which feels more natural
A source than that of a furnace fed
With oil. We justify our use: trees are renewable.
But too many homes reliant have denuded

Whole landscapes, engendered injurious air.
What choice? We are here. Our lifestyle we pledge to spare.

Combustible

Create for me mortals, said Zeus, and I'll spare
You imprisonment. It's one version.
The rest is undisputed: the dare
Prometheus took to defy his Father, son

Of a Titan. So humans were given
Fire. There's back and forth, and trickery,
The young God bound to rock, the eagle, the riven
Liver. Travel along, from myth to history:

How we learn to ignite the flame, from sticks
To hand drill to flint & steel, then the match, finally,
Whose most salable permutation meant coating the tips
With white phosphorous, an element so easily

Combustible that air alone makes it blaze.
Nothing new to tell: be careful what you praise.

Fashioning the Match

It isn't animal, of course, or even louse that preys
On us, but the shilling and meager pence. Here's a dirge
For the factory workers, mostly women, whose days
Were spent in routine toil at a ghastly wage

Fashioning the match. First the splints of poplar or pine,
Each end dipped in a chemical blend, transferred
To ovens, then halved and packed, 100 per box. A fine
For the lass with an untidy bench, who exchanged a word

With her neighbor, or whose soiled feet—bare,
Because shoes were dear—offended the principled
Quakers. Bryant & May, sprung in East London, its share
For stockholders 20 percent, while the Mick who complained

Of a toothache was typically sacked. Death or disfigurement
Might follow. A mandibular horror no author could invent.

Blind Trap

Charles Sauria, French chemist, first to invent
The phosphorous match. Non-explosive, non-
Odorous, flame constant: a notable improvement
To the Lucifer. Alas, poor hireling—son

Or daughter of charwoman or pauper,
The factory was your default: a blind trap
Where the human jaw collected toxic vapor
Swelled and decayed, configured a map

Of pumice-like bone, erupted with foul pus.
Consider: Mrs. Fleet. Four teeth plucked,
Part of her mandible lost, the smell so gross
Her family recoiled. Or John Werner, chin tucked

In a bandage knotted atop his head, his nourishment
All liquid. Ashland, Ohio: another immigrant spent.

Halo Effect

For every factory owner who spent
A single sleepless night plagued by the fate
Of his workers, the odds are a hundred didn't.
They knew the ailments, but the exchange rate

Was easy: one employee equaled another
Thousand. Therefore, let the sick ones go,
Deny wrongdoing in any lawsuit: improper
Conduct, negligence, testimonies of the glow

Emitted from a stricken jaw. Yes,
In the dark, one might see it: a halo effect,
Greenish-white, not unlike bioluminescence—
Lost spirits or fairies? An eye socket

Missing, nausea, vomiting, convulsions;
The organs' arrest, for the lucky ones.

Fissure

When fire first moved from terrain to hearth is anyone's
Guess. Control, the shift suggests, the skewered bird
Or leg of ungulate sizzling on hot stones
As smoke in the crude chimney tunneled upward.

Devil fire, in orange guise, dancing a solo tango—
Leap and pirouette mask your true intent:
To find a fissure in clay or rock, to follow
It out, flee the grip of your strange containment.

Then—to consume the hovel, the life within,
Return to grass and wood, acre upon acre—
As today, you sweep Australia, win
Back your sovereignty, your rank as Master.

Koala, kangaroo, bird, bat, frog… statistics blur.
One billion lost. "It could be a very conservative figure."

Cave Wall to Canvas

Art is replete with the human figure.
From cave wall to canvas, our image is sketched
Sometimes in a manner less than laudatory. A mirror
We'd rather eschew, like the 80 plates etched

By Goya, *Los Caprichos,* which condemn
Human foolishness and folly. Or half-beasts
Depicting our gluttony and lust, Rubens'
Satyrs, Picasso's minotaurs, along with feasts

Of Dionysian indulgence. Bathers, lovers,
Dancers... The Sleeping Venus... settle the eye,
Presume a mythic innocence. The future's
Sure. The wailing babe is rocked and sung a lullaby

While half the planet erupts in flame. *Quiet, child. Lay*
Down your head, Turn, and turn, and turn, and turn away.

Call it Romance

Bright petals of rain are falling away
From branch of willow and birch. A skiff of snow
Last night lay frozen until dawn; today
The melt will leave us mourning. Winters ago

White would wrap our shoulders like a shawl;
The sound of our own step, muffled, diminished
Our importance in the chill expanse. Call
It romance, enchantment with the past; what's extinguished

Is a lamp internal, a remembrance that I'll never live
To light again: relinquishment to another order, a sense
That we, dear sapiens, could manage to forgive
Our trespasses, that Nature would earn her recompense.

Now the trees are bare, their limbs dark fingers of admonition:
Every blackened match head is a marker of our own destruction.

Augury

It is past time for the destruction
Of error, since the first wise monkey descended
The first tree, since the first perfect stick was broken
To become a probe for honey, or a spear that ended

A brother's life. Or fuel for the fickle new companion
Whose charm of warmth was a brilliant malevolent
Ploy to draw the willing body near, then turn
And leap—a flame insatiable and silent—

To singe the simian's skin and leave him howling.
He did not learn. He did not comprehend
The augury, the burn that would keep on blossoming
Through all of history, the land that would not mend.

Peat, dung, coal, diesel, gasoline. The oven, the smelter.
What cost, this comfort; how delirious, this fever.

Swab

I dream I recover from a long fever.
Shiver and sweat, ache and flush are finished.
The tongues of hell have turned to lapping water.
I'm a diatom, a nautilus—buoyant, replenished,

Floating in a light sea, the color musically cool,
White azure. This is the world that hangs on the edge
Of consciousness, when I wake, and a new rule
Seems possible… a planetary pledge

To reduce the temperature, lay the moist cloth
On the globe, as one would swab the forehead
Of a sick child. To rise, healed, from this time forth,
Together—no longer conversant with the dead.

Is this unreal? A dream so strong can for hours linger.
O let it, for a day or three, or a week, endure.

Gullible Hope

Gullible Hope has managed to endure
Since Pandora first trapped the spirit in her jar,
While evil—outside—thrived. A firefly so kept is sure
To die, and, with our current climatic war

Even the ancient Greeks would have marveled
At Her survival. Though whether Elpis was deemed
A gift or curse was long debated. For the grizzled
Norsemen, Vön was poorly rated: slobber that streamed

From the jaws of Fenris Wolf, verse less than suitable
For the Bible. While Greta, our solemn Swedish maiden,
Is partial to Panic. "I don't want you to be hopeful,"
She declares. She flips back her braids and urges action.

I look to the sky. No moon tonight, and I've lost
Her last position. She'll wax again and wane, before Earth's Holocaust.

Holos

Holos 'whole' with *kaustos* 'burnt' is holocaust—
For ancients, sacrifice by fire, burnt offerings.
Moderns present them via car and ship exhaust:
Toxic gases, particles that clog the lungs. O Earthlings!

Forever giving—no mere bowl—but a heavenful
Of benefactions. Imagine what's up there:
Fumes of goat, sheep, bullock, pigeon and turtle-
Dove. All the heated breath that's fashioned into prayer.

Not to mention smoke from the Amazon
And Indonesia. There's more, of course, but the list
Is tedious; God is stifling a wide yawn.
Enough! You'd think by now he would insist

We curb the burning. We need a sign: monsoon and hurricane
That scald and blister the skin: a blazing wind, a searing rain.

Soot and Goo

Who, left with the bones of the last charred mammal, will reign?
The mad painter may appear the victor, the surface of Earth
Rendered in gray and black. Charcoal, ash: all that will remain
Of the human chronicle. Our petty souvenirs, our worth

Spread like dirty smears on glass. Our achievements here
Forgotten, while fire, not to be outdone, burns covertly
Underground. Never discount the chandelier
In the coal seam that can blaze for an eternity

Or the Nematodes living three thousand feet down.
They need less oxygen and thrive in heat. Insect
And worm will rise up stronger when we're gone,
Tooling around in the soot and goo, an art project

To rival Picasso's best. Creation doesn't cease; the world
Births multiple masters, and will, in our aftermath, be filled.

Utterance

A new canvas, according to Da Vinci, should be filled
With a wash of black, while Matisse, wild beast, would allow
Bare weave in finished work: a plot of canvas left untilled.
I take up my brush today resolved to follow

The French rebel, for the mountains of my home
Are best left white. There's snow again this morning
On the near peaks, as if the beauty of untrammeled form
Insists on jagged cuts that surge, stark and unforgiving,

From ice and rock. I too can be recalcitrant
Even as I keep on painting. To live is to create
And thus destroy. But I vow not to cover over, to grant
This picture some scant relief, knowing it's too late

For remedy. Our tenure here might have deserved some utterance
But for the many forms of life that perished at our expense.

Notes

"Veteran Creator," p.10

Begin, he says, with madder brown/ For alder ... if we cannot halt / This growth ... a tangled pattern consuming track/ Of caribou;

Climate change in parts of Alaska has resulted in "a widespread increase in alder abundance and distribution"

Clark, Robert, et al. "The Effects of a Changing Climate on Key Habitats in Alaska." Alaska Dept. of Fish and Game, September 2010. <https://www.adfg.alaska.gov/static/lands/ecosystems/pdfs/sp10_14.pdf>

"Cast in Steel," p.12

Bristol Bay fishing boats no longer manned by sail.

The year 1952 was pivotal for the commercial fishing industry of Bristol Bay. That was the year a federal regulation preventing the use of motorized vessels was repealed, a move that sounded the death knell for the double-ender sailboats.

Brown, Tricia. "Bristol Bay Sailboats." Lit Site Alaska. N.d. <http:/www.litsitealaska.org/index.cfm?section=digital-archives &page=Land-Sea-Air&cat=Ships-and-Boats&viewpost=2&ContentId=2644>

"Her Loud Red Coat," p.13

And that is why Jane, in her loud red coat/ Marches on Washington ... / her best leading part/ Since '78./ A kind of homecoming for her,/ Protest of a different war: this one global

Jane Fonda has moved her star-filled climate change protests from Washington, D.C. to Los Angeles. At the first rally in the new location, the actress, 82, was joined by...

Moniuszko, Sara M. "Jane Fonda moves climate change protests to Los Angeles, rallies with Joaquin Phoenix, more stars." USA Today, February 7, 2020. <https://www.usatoday.com/story/entertainment/celebrities/2020 /02/07/jane-fonda-joaquin-phoenix-protest-climate-change-los-angeles/4692793002>

Jane Fonda won her second "Best Actress" Oscar in 1978 for the anti-Vietnam war film, *Coming Home.*

Levy, Emanuel. "Oscar: Fonda, Jane in Coming Home (1978)." Cinema 24/7, January 10, 2011. <https://emanuellevy.com/oscar/oscar-history-best-actress-jane-fonda-in-coming-home-1978-6>

"Spell of Pigment," p.15

Still as stones or wool

From "Walking Around" by Pablo Neruda (Robert Bly translation): *The only thing I want is to lie still like stones or wool.*

"Yarn," p.16

And chimney sweeps—young boys, five or eight/ Years of age—heads shaved so they wouldn't catch/ Fire, the fine soot that infused their lungs and plagued/ Their skin with cancer

See full text of "The Chimney Sweeper" (*Songs of Innocence*) by William Blake

"Combustible," p.18

then the match, finally,/ Whose most salable permutation meant coating the tips/ With white phosphorous

White phosphorus is the most reactive, the least stable, the most volatile, the least dense, and the most toxic of the allotropes.

Lancashire, Robert J., Professor. "Phosphorous and the History of the Match," February 2014. Many of these course notes were sourced from Wikipedia under the Creative Commons License. <http://wwwchem.uwimona.edu.jm/courses/CHEM2402/Myths_General/P_Matches.html>

"Fashioning the Match," p.19

the principled/ Quakers. Bryant & May, sprung in East London

The Bryant and May match factory had a large workforce that included over 1400 women and girls.

A mandibular horror

Phossy jaw: Resulting from chronic phosphorus poisoning, the condition in the initial stages causes dental pain which can be quite difficult to bear. In most cases, the lower jawbone is affected.

"Facts About Phossy Jaw: A Deadly Occupational Disease" <https://healthhearty.com/facts-about-phossy-jaw-deadly-occupational-disease>

See also:

Brain, Jessica. "The Match Girls Strike." Historic UK, April 2021. <https://www.historic-uk.com/HistoryUK/HistoryofBritain Match-Girls-Strike/>

"Blind Trap," p.20

The phosphorous match. Non-explosive, non-/ Odorous, flame constant: a notable improvement/ To the Lucifer.

Samuel Jones patented the *Lucifer* match. These early matches had a number of problems—an initial violent reaction, an unsteady flame and unpleasant odor and fumes. Lucifers could ignite explosively, sometimes throwing sparks a considerable distance.

White phosphorus matches were theoretically safer but ensuing health problems for factory workers proved otherwise.

Beller, Misty M. "The History of Matches." Heroes, Heroines, and History, July 2020.<https://www.hhistory.com/2020/07/the-history-of-matches.html>

Mrs. Fleet. Four teeth plucked,/ Part of her mandible lost

There were no medications to treat phossy jaw. Treatment involved surgically extracting the jawbone.

John Werner

John Werner came to America from Germany and worked in an Ohio match factory for just over a year, beginning in August 1907. White phosphorus match production had been banned in most civilized countries, but not in the U.S. Jobless, homeless, and maimed for life, Werner tried in vain to raise enough money to return to Germany, where his wife and daughter had relocated.

Andrews, John B. "A Terrible Story: Matchworker Contracts Disease Without Hope of Recovery" Trove. Originally published in the Survey. Reprinted in the Daily Herald, February 1912. <https://trove.nla.gov.au/newspaper/article/105214869>

"Halo Effect," p.21

the glow/ Emitted from a stricken jaw.

It is said that the infected bones even glowed in the dark, giving off a greenish-white halo effect.

"Facts About Phossy Jaw: A Deadly Occupational Disease" <https://healthhearty.com/facts-about-phossy-jaw-deadly-occupational-disease>

"Fissure," p.22

As today, you sweep Australia

Known colloquially as the *black summer*, the fires burned an estimated 46 million acres, from June 2019 into March 2020.

<https://en.wikipedia.org/wiki/2019%E2%80%9320_Australian_bushfire_season>

"Augury," p.25

It is past time for the destruction/ Of error

A reworking of line one of W. H. Auden's "The Destruction of Error": *It is time for the destruction of error.*

"Swab," p.26

The tongues of hell

From "Fever 103°" by Sylvia Plath

"Gullible Hope," p.27

Gullible Hope

In the Timaeus, Plato also adopts a rather negative attitude towards hope by recounting a myth according to which the divine beings give us "those mindless advisers confidence and fear, (...) and gullible hope."

Bloeser, Claudia. "Hope." Stanford Encyclopedia of Philosophy, First published Mar 8, 2017, revised March 2022. <https://plato.stanford.edu/entries/hope>

Elpis (Hope) appears in ancient Greek mythology with the story of Zeus and Prometheus. Vön (Hope): Norse mythology however considered Hope (Vön) to be the slobber dripping from the mouth of Fenris Wolf, their concept of courage rated most highly a cheerful bravery in the absence of hope.

Anonymous, "Hope." May 2016. <https://astudenthope.blogspot.com/search?q=slobber>

Greta, our solemn Swedish maiden

Greta Thunberg is a Swedish environmental activist on climate change whose campaigning has gained international recognition. In August 2018, at age 15, she started spending her school days outside the Swedish parliament to call for stronger action on climate change by holding up a sign reading Skolstrejk för klimatet (School strike for climate).

<https://en.wikipedia.org/wiki/Greta_Thunberg>

"Holos," p.28

We need a sign: monsoon and hurricane/ That scald and blister the skin: a blazing wind, a searing rain

Jewish legend: God bade each drop (of water) pass through Gehenna before it fell to earth, and the hot rain scalded the skin of the sinners. The punishment that overtook them was befitting their crime. As their sensual desires had made them hot, and inflamed them to immoral excesses, so they were chastised by means of heated water (Ginzberg, 1909, 1:106, emp. added).

Miller, Dave. "The Quran and the Flood." Apologetics Press, April 2012. <http://www.apologeticspress.org/APContent aspx?article=3509>

"Utterance," p.30

A new canvas, according to Da Vinci, should be filled/ With a wash of black

"A painter should begin every canvas with a wash of black, because all things in nature are dark except where exposed by the light."
—Leonardo da Vinci

Matisse, wild beast

The name *les fauves* ('the wild beasts') was coined by the critic Louis Vauxcelles when he saw the work of Henri Matisse and André Derain in an exhibition, *the salon d'automne* in Paris, in 1905. Fauvism emphasized painterly qualities and strong color over the representational or realistic values retained by Impressionism.

"Fauvism" <https://www.tate.org.uk/art/art-terms/f/fauvism>

Acknowledgments

Gratitude to the editors of the following literary journals, where a few of the poems from this collection were originally published:

Alaska Quarterly Review: "Blind Trap," "Gullible Hope," and "Holos"

Kestrel: A Journal of Literature and Art: "First Infraction," "Call it Romance," and "Augury"

Merci mille fois to fellow poet Ruth Holzer for her sharp eyes and smart edits.

Praise for *Late Fall Bucolics*

I've been an admirer of Anne Coray's tough, lively nature poems for many years. In *Late Fall Bucolics* the natural world again takes center stage, a planet especially raw, turbulent, and angry, as if lashing out in its own last defense. These poems chart an elemental storm of fire and ice, of seasons out of whack, a terrain under siege by human ignorance. "All will burn, but how magnificent the color." Woven throughout is a complementary examination of landscape painting (by amateur and master alike)—the inadequacy of art's mimicry offset by the compulsion to witness, to fix on canvas some testimony to the terrible beauty that is quickly and forever passing. Ms. Coray seems energized by the parameters and possibilities of the sonnet in this linked sequence, and despite her contention that it is "too late/ For remedy," the consistent flashes of play here, the continual linguistic energy, and most centrally the poet's enduring gaze—even at her own culpability—create a voice urgent and desirous, perhaps even hopeful, that "something remains of place."

—Gaylord Brewer, author of *Worship the Pig*

Late Fall Bucolics ranges through space and time, looking out toward peaks and back toward industrial history, connecting wool mittens in the closet with the shaved heads of chimney sweeps to ever-quickening looms to the "modern blanket" of our warmed planet. Sonnets and bucolics are both satisfying forms: recognizable modes of making music that add another layer of resonance when a reader considers their origins in relation to contemporary use. What does Stratford-upon-Avon or Ancient Greece have in common with Western Alaska? Well… love, patterning, attunement to the cycles of the more-than-human world. In Anne Coray's sonnets, loss of a known pattern of seasons is as heartbreaking as any other love lost. *Late Fall Bucolics* thrums with art, fire, warmth, the embers of autumn-upon-winter in the hearth and landscape—and how to grapple with our place in the Anthropocene.

—Elizabeth Bradfield, author of *Toward Antarctica*

About the Author

Anne Coray is the author of the novel *Lost Mountain* (West Margin Press) as well as three poetry collections—*Bone Strings, A Measure's Hush,* and *Violet Transparent.* She is also the coauthor of *Crosscurrents North: Alaskans on the Environment.*

The recipient of fellowships from the Alaska State Council on the Arts and the Rasmuson Foundation, she divides her time between her birthplace on remote Lake Clark (*Qizhjeh Vena*) and the coastal town of Homer, Alaska.

About The Poetry Box®

The Poetry Box,® a boutique publishing company in Portland, Oregon, provides a platform for both established and emerging poets to share their words with the world through beautiful printed books and chapbooks.

Feel free to visit the online bookstore (thePoetryBox.com), where you'll find more titles including:

The Catalog of Small Contentments by Carolyn Martin

Earth Resonance: Poems for a Viable Future by Sam Love

A Shape of Sky by Cathy Cain

A Long, Wide Stretch of Calm by Melanie Green

Of the Forest by Linda Ferguson

Let's Hear It for the Horses by Tricia Knoll

Stronger Than the Current by Mark Thalman

Sophia & Mister Walter Whitman by Penelope Scambly Schott

A Nest in the Heart by Vivienne Popperl

What We Bring Home by Susan Coultrap McQuin

In the Jaguar's House by Debbie Hall

Beneath the Gravel Weight of Stars by Mimi German

Tell Her Yes by Ann Farley

and more . . .

CPSIA information can be obtained
at www.ICGtesting.com
Printed in the USA
JSHW021925290622
27453JS00001B/49